Companion Guide to

The Promises and Perils of

AI in Education

I0530386

KEN SHELTON and DEE LANIER

"Can they prove to me that whatever AI they have is actually doing a better job than the simplest possible solution that I have?"

-Dr. Timnit Gebru

Table of Contents

INTRODUCTION TO COMPANION

The purpose of this companion guide is to support your journey through our book, *The Promises and Perils of AI in Education*, by providing structured opportunities to deepen your understanding of the book's key concepts, reflect on the challenges posed by AI in educational contexts, and apply what you've learned to real-world scenarios. We recognize that educators across all dimensions of responsibility may be at different levels and paths, so we designed this guide to be valid regardless of where you are. The goal here is not to compare yourself with others but to gauge where you currently are and where you will be upon completing the activities in the guide.

Organized systematically, the guide is structured in a uniform format where each chapter section contains three consistent segments:

Concepts: Each chapter of this companion guide begins with concepts related to the chapter's central themes, presented with guiding questions to help you reflect, make connections, and frame discussions with friends, colleagues, and your book study group. Beyond the key terms, we encourage

1

you to explore ideas that may be less familiar. For instance, Chapter 5 introduces the concept of design thinking. You might explore different design thinking approaches in your group discussions, such as the Stanford d.school method, Liberatory Design, or Solve in Time. Reflect on and discuss these and any other new concepts. Consider how they relate to the ethical evaluation and equitable use of AI, as discussed in each chapter.

Problems Identified: Each chapter explores the potential perils of AI, particularly its role in automating inequities, prompting critical thinking about its impact. Often, these issues are masked behind surface-level platitudes such as "tips and tricks" or "top ten" lists, which trivialize them as puzzles—intellectual challenges designed to test ingenuity without bearing real-life stakes. However, we label these as problems because they are real issues that cause stress, suffering, and harm to individuals and their communities. They are not merely abstract concerns but have real-world impactful consequences. Unaddressed, they will have dire consequences, particularly for our most vulnerable populations. As you review each chapter, you may identify even more problems. We urge you to sit with them, reflect on them, discuss them, and collaborate with your community to find ways to overcome/solve them. Leaving these problems idle and being consciously complicit will result in definite peril. Your initiative and actions promise a more hopeful future with AI.

Tasks: Practical activities designed to help you explore solutions, engage in creative thinking, and apply your learning meaningfully. The tasks will include a guide to prompting protocols for you to utilize for each chapter, along with additional tasks where you will be able to synthesize and apply your learning to real-world scenarios.

Note: The prompts in the task section are designed to enhance your exploration of the concepts and problems in each chapter. While each guide aligns with a specific chapter, feel free to adapt and apply them across other chapters. In fact, we encourage this exploration. They are designed as flexible approaches to inspire your thinking, not as strict templates. Many of these prompts work effectively beyond the chapter they're paired with, offering a versatile approach to deepening your insights throughout the book.

Introduction to thinking routines and their role in reflection

Thinking routines are structured practices that encourage reflective thinking and deeper engagement with the material. In this companion guide, thinking routines will help you absorb, critically analyze, and connect information to your experiences. Some routines you'll encounter include:

See, Think, Wonder: What do you notice (see)? What do you think is happening (think)? What do you still think about (wonder)?

Connect, Extend, Challenge: How does this connect to what you already know? How does this extend your thinking? What challenges or questions does it raise for you?

These routines are intended to spark curiosity and reflection, enabling you to develop a deeper, more nuanced understanding of AI's role in education.

Importance of digital literacy, digital fluency, AI literacy, and ownership of learning

Digital literacy and digital fluency are essential foundational skills for navigating the complexities of AI and technology in education within a contemporary, relevant, and responsive context. Digital literacy refers to the ability to find, evaluate, and use digital information effectively. Digital fluency goes beyond literacy; it involves creating digital content, solving problems, and applying critical thinking to new and emerging technologies. The foundation of AI literacy rests on building practical skills, comprehending AI systems, and developing sound judgment when assessing AI platforms—essential elements for thriving in an AI-driven world.

This guide encourages you to take ownership of your learning journey. By actively engaging with the material, questioning assumptions, and applying what you've learned to your context, you will develop the fluency needed to navigate AI tools ethically, critically, and responsibly.

Key reminders such as "More Context = Better Output"

Remember that context matters as you work through the tasks in this guide. Whether you're reflecting on how AI is used in your educational environment or brainstorming solutions to a problem, the more context (present, historical, and social) you provide, the better your insights and outcomes will be. AI systems rely on context to provide accurate and relevant responses, and the same principle applies to your thinking process—more information, more perspectives, and more reflection lead to more robust understanding and higher effective results.

Encouragement to do this work with others

Collaboration is the key to unlocking a deeper understanding of the complex landscape of AI in education. We encourage you to engage with your peers, colleagues, or fellow educators, particularly within book study groups, as you journey through this Companion Guide.

By discussing the challenges, opportunities, promises, and perils of AI in education with others, you can gain valuable insights and foster a more collaborative approach. Share your thoughts, ask questions, and listen to the perspectives of others. Together, we can navigate the ethical and equitable implications of AI in education.

Remember, this is a collective effort. By working together, we can ensure that AI is used responsibly and effectively to enhance learning for all.

More room to capture additional notes

The Companion Guide, much like its original counterpart, invites you to engage actively with the text. Its wide margins provide ample space for personal notes, questions, doodles, and reflections. Whether you connect ideas to your own experiences, visualize complex concepts, or simply express your thoughts, these margins are your canvas. Dot journal pages at the end of each chapter offer a creative space for mind mapping, brainstorming, or artistic expression. Experiment with page orientation, colors, and styles to personalize your learning journey.

Chapter 1 Concepts

Review the following key concepts in Chapter 1: *AI, Machine Learning, Deep Learning,* and *Natural Language Processing.*

1. Identify and define any additional terms or concepts from this chapter that were new to you.

2. Reflect on your notes from this chapter in the book. Which concepts or ideas would you like to discuss further?

3. How do AI, machine learning, deep learning, and natural language processing concepts relate to education?

Chapter 1 Problems

Reflect on the following potential *perils* of AI in education and discuss them with others:

1. **Bias in AI Educational Tools:** AI educational tools can perpetuate biases that already exist in society. For example, facial recognition software used in schools may disproportionately misidentify students of color, leading to unfair treatment and discrimination. Using an AI system for enrollment or scheduling purposes may disproportionately affect students who have been historically excluded from specific programs or classes.

2. **Student Access to AI Tools:** Schools often restrict student access to AI tools while allowing teachers to use them. This creates an uneven playing field and limits students' opportunities to develop digital skills. Schools, districts, boards, or school systems may create "guidance" on using AI using oversimplified criteria like stop (red), slow yellow), and go (green) or some visual

equivalent, leaving the decision-making solely with the teacher.

3. **Unreliable Plagiarism Detection:** AI plagiarism detectors are not always accurate and can flag students' work as plagiarized even if it is not. This can lead to unnecessary stress and anxiety for students. AI detectors also show a statistically noticeable bias against emerging multilingual learners acquiring English.

4. **Privacy Concerns:** The use of AI in schools raises privacy concerns. Students may not be aware of how their data is being collected and used by AI systems. Teachers and administrators may not be aware of the data practices employed by AI developers.

5. **Transparency in Design:** The algorithms used in AI educational tools are often opaque, hindering accountability for biases or errors. This lack of transparency makes it difficult to understand how the tools work and their impact on students.

Chapter 1 Tasks

Tasks:

Using the provided LLM prompt, create a dialogue between a teacher and a student discussing AI in education.

Persona

1. **Give** it a **persona**
   Act as a [specific role or profession].

2. **Give** it a **purpose**
   Your goal is to [clearly state the objective].

3. **Give** it **parameters**
   Consider these factors: [list relevant constraints, guidelines, or preferences].

4. **Polish** your results
   If necessary, I will provide feedback and further instructions to help you refine your output.

Sample Prompt (reminder, the samples throughout this guide are examples but by no means a prescriptive approach): Act as a high school teacher who is knowledgeable about technology and enthusiastic about explaining complex ideas in a simple, relatable way. Your goal is to engage a curious student who wants to know more about how AI is used in their school and what it might mean for their future. Consider these factors:

- The student may have limited prior knowledge about AI.
- Use everyday examples to clarify concepts like "machine learning" or "data privacy."
- Encourage the student to ask questions and express any concerns.

What do you notice and wonder about the results?

What connections can you make to this chapter?

Additional Task (Scenario Analysis):
At ABC High School, teachers are starting to use an AI-powered tool that offers personalized learning recommendations based on students' previous performance. However, the AI system is only accessible to teachers and not directly to students. Teachers are responsible for using the tool to guide their instruction, but students often need to be made aware of the AI's role in shaping their learning paths.

A group of students has started questioning why they aren't able to access the same AI tools their teachers are using, especially as they feel it might help them develop their skills in self-directed learning and increase their engagement. Some students argue that having access would empower them to understand their progress better and take more control over their education. However, the school administration is concerned about how students might misuse or misunderstand AI's recommendations.

At the same time, one student raises concerns about the tool's fairness. They have noticed that the AI recommends more advanced content and materials to students from wealthier backgrounds, as they tend to have higher previous scores. In contrast, students from less advantaged backgrounds receive more remedial content and material recommendations. The student wonders if the AI is perpetuating inequities by reinforcing pre-existing gaps in student performance.

Analyze the situation described in the scenario and identify how the problem of student access to AI tools might be impacting equity and learning in the classroom.

Discuss what the students could ask or might recommend to ensure the AI system is being used fairly. How could the school ensure that all students, regardless of background, have equal opportunities to benefit from AI?

Chapter 2: Examining Bias in Educational AI

Chapter 2 Concepts

Review the following key concepts in Chapter 2:
Choice-supportive Bias, Bandwagon Effect,
Selective Bias, Marginalization, and Rightsholders.

1. Identify and define any additional terms or
    concepts from this chapter that were new to
    you.

2. Reflect on your notes from this chapter in
    the book. Which concepts or ideas would
    you like to discuss further?

3. How do the concepts of choice-supportive
    bias, bandwagon effect, selective bias,
    marginalization, and rightsholders relate to
    AI in schools?

Chapter 2 Problems

Reflect on the following potential *perils* of AI in education and discuss them with others:

Transparency in Data Acquisition and Algorithmic Development: Chapter 2 highlights the lack of transparency in how data is collected and used to train AI systems. This can lead to biased data inclusion during training, which the AI itself can then perpetuate.

Limited Oversight and Accountability: Chapter 2 points out the limited oversight and accountability for AI development in education. This can make it challenging to identify and address bias in AI systems.

Perpetuation of Stereotypes: Chapter 2 discusses how AI systems can perpetuate stereotypes about certain groups of students. For example, an AI system used to identify students with learning disabilities might be more likely to flag students from marginalized backgrounds.

Exacerbation of the School-to-Prison Pipeline: Chapter 2 emphasizes how AI systems could exacerbate the school-to-prison pipeline by recommending harsher disciplinary actions for students from specific backgrounds.

Limited Exposure to Diverse AI Tools: Chapter 2 mentions limited exposure to diverse AI tools designed to reduce bias. This can lead to biased AI systems becoming the norm in education.

Chapter 2 Tasks

Tasks:

Using the provided prompt, create a dialogue between you and an AI LLM.

Interactive

1. **Creating the Dialogue**
   I want you to engage me in a dialogue about [clearly state the topic].

2. **Further modification**
   Your goal is to ask me questions that will [explain the desired outcome of the dialogue, e.g., help me explore this topic, make a decision, solve a problem, etc.].

3. **Additional parameters**
   Ask one question at a time, and wait for my response before proceeding.

Sample Prompt: I want you to engage me in a dialogue about identifying and addressing bias in educational AI systems. Your goal is to ask me questions that will help me explore how bias may appear in AI tools used in schools, consider its impacts on students, and think critically about ways to mitigate it. Ask one question at a time, and wait for my response before proceeding.

What do you notice and wonder about the results?

What connections can you make to this chapter?

Additional Task (Scenario Analysis):
The Valley Middle School recently adopted an AI-powered discipline monitoring system to track behavioral incidents and recommend appropriate student consequences. The system analyzes data from school-installed cameras, student behavior records, and teacher reports to predict potential disruptions and suggest interventions. Teachers have reported that the AI seems effective in helping them manage the classroom and maintain order.

However, one of the teachers starts noticing a disturbing trend. The AI system disproportionately flags students of color, mainly Black and Hispanic students, as being at risk for behavioral issues, even in cases where their behavior seems similar to their white peers. This teacher raises his concerns with the administration, but they argue that AI simply follows data patterns. Worried that the AI might be reinforcing racial stereotypes and unfairly penalizing students of color, the teacher decides to investigate further. The teacher discovers that the historical data the AI was trained on includes disproportionate disciplinary actions against students of color, and the AI's recommendations perpetuate this bias.

This bias in the AI system can have serious consequences for Black and Brown students. They may be unfairly punished, labeled as troublemakers, and denied opportunities for learning and growth. Additionally, such biased AI systems can erode trust between students of color and school authorities, further exacerbating the problem.

Analyze the situation described in this scenario. Identify how bias has influenced the AI system. What are the potential, even unintended, consequences for students if this bias is not addressed appropriately?

Discuss the steps the teacher could take to advocate for fairness when using this AI tool. What kind of data transparency should be demanded from the system?

Finally, consider the broader implications of AI bias in education. How can educators and policymakers work together to ensure that AI tools are developed and used ethically and equitably?

Chapter 3: The Digital Divide in Schools

Chapter 3 Concepts

Review the following key concepts in Chapter 3: *Digital Divide, Digital Equity, AI Literacy, and Digital Fluency.*

1. Identify and define any additional terms or concepts from this chapter that were new to you.

2. Reflect on your notes from this chapter in the book. Which concepts or ideas would you like to discuss further?

3. How do the digital divide, digital equity, AI literacy, and digital fluency concepts relate to the students in your school(s)?

Chapter 3 Problems

Reflect on the following potential *perils* of AI in education and discuss them with others:

Diverse Representation in AI Development: Chapter 3 highlights the lack of diverse representation in AI development. This can lead to AI systems that reflect and perpetuate the biases of the programmers who create them. For example, if white men primarily develop AI systems, they may be more likely to disadvantage students from other backgrounds.

Limited Exposure to AI for Minority Students: Chapter 3 discusses limited exposure to AI for minority students. This can exacerbate existing achievement gaps and limit opportunities for these students in the future workforce. Students not exposed to AI in school may be less prepared for jobs requiring AI skills.

Perpetuation of Stereotypes in AI Learning Tools: Chapter 3 points out the perpetuation of stereotypes in AI learning tools. AI learning tools that are not designed carefully can reinforce negative stereotypes about certain groups of students. For example, an AI tutor that is more likely to provide positive feedback to students from wealthy backgrounds could perpetuate stereotypes about intelligence and socioeconomic status.

Culturally Relevant AI Tools: Chapter 3 emphasizes the need for culturally relevant AI tools. These tools can help students from diverse backgrounds see themselves reflected in technology and feel more engaged in learning. Without access to culturally relevant AI tools, students from marginalized groups may feel alienated from AI and less likely to pursue careers in AI fields.

Limited Training for Teachers on AI Bias: Chapter 3 mentions the limited training for teachers on AI bias. Teachers must know the potential for bias in AI systems and how to mitigate it. Without proper training, teachers may be unable to identify and address bias in the AI tools they use in their classrooms.

Tasks:

Using the provided prompt, create a comparative analysis between two different schools analyzing equity of access on AI in education.

Comparative

1. **Provide** two differing **situations or conditions**
   Within the scope of [specific area or topic], compare and contrast [what you want to compare in relation to the scenarios], focusing on [pros and cons] or [conditions of each].

2. **Expanding the analysis**
   To help me fully understand the comparison, ask me clarifying questions about [specific aspects relevant to the scenarios]. Then, suggest different approaches I could take for [desired outcome in each scenario].

Sample Prompt: Within the scope of the digital divide in educational AI access, compare and contrast a well-funded suburban school with ample AI resources versus a low-income urban school with limited access to AI tools. Focus on each school's benefits and challenges in offering AI-supported learning experiences. To help me fully understand the comparison, ask me clarifying questions about the specific impacts of AI resource availability on students' engagement and learning outcomes. Then, suggest different approaches I could take to improve AI accessibility and equity in each scenario.

What do you notice and wonder about the results?

What connections can you make to this chapter?

Additional Task (Scenario Analysis):
In two neighboring school districts (this scenario may also occur in the same district but two different schools), Mountain Valley School District and City School District, AI-powered creative tools are being introduced to enhance student learning. Mountain Valley School District, a well-funded suburban district, has equipped each student with personal laptops and access to a suite of AI-powered tools for creative projects.

These tools allow students to generate artwork, music, and even assist with coding projects, enabling them to explore their creative potential with cutting-edge technology. Meanwhile, City School District, a lower-income, primarily urban district, struggles with some outdated equipment and limited funding. Students are provided with computers in a 1:1 environment. They often encounter software and usage restrictions, and the district believes that the students cannot use any programs until after the teachers reach a full comfort level first. As a result, this prevents them from using the same AI-powered tools available at Mountain Valley. Teachers in City School District work hard to make up for the technological gaps.

Still, many students feel left behind, especially when they hear about the advanced projects their peers in Mountain Valley are completing. One student in City School District is passionate about music production and dreams of using AI tools to create compositions, but the lack of access to these tools limits their ability to explore this passion. This student starts to wonder whether this technology divide affects creativity and long-term opportunities for students from less privileged backgrounds.

Analyze how the digital divide between the two districts impacts students' ability to access and benefit from AI-powered creative tools. How might this divide widen existing educational inequities? Opportunities?

Discuss what steps the City School District could take to provide more equitable access to AI tools. What role can educators, policymakers, or communities play in addressing the digital divide?

Chapter 4: AI and High-Stakes Decisions in Schools

Chapter 4 Concepts

Review the following key concepts in Chapter 4: *Neural Networks, Data Weighting, Data Cleansing, and Algorithmic Auditing.*

1. Identify and define any additional terms or concepts from this chapter that were new to you.

2. Reflect on your notes from this chapter in the book. Which concepts or ideas would you like to discuss further?

3. How do neural networks, data weighting, data cleansing, and algorithmic auditing concepts relate to ensuring fairness and equity in educational systems?

Chapter 4 Problems

Reflect on the following potential *perils* of AI in education and discuss them with others:

Transparency in AI Assessment Systems: Chapter 4 emphasizes the lack of transparency in AI systems used for educational assessment. This makes it difficult to understand how these systems function and make decisions. Without transparency, holding these systems accountable and ensuring fairness is challenging.

Potential for Bias: AI systems can perpetuate biases in the data they are trained on. For example, an AI system used to grade essays may be biased against certain writing styles or topics, leading to unfair student outcomes.

Privacy Concerns: The use of AI in education can raise privacy concerns. For example, AI systems may collect data on students' personal lives, such as browsing history or social media activity. This data could be used to track students or target them with advertising. Or an AI system may fully comply with data privacy laws yet still utilize user data to further train its model without user knowledge or consent.

Overreliance on AI: There is a risk that educators may become too reliant on AI systems. AI should be used to support educators, not replace them. Human judgment is still essential in education.

Limited Access to Resources: Not all schools have the resources to invest in AI systems. This could create an equity gap, where students in wealthier schools have access to AI-powered educational tools that poorer schools do not.

Tasks:

Using the provided prompt, create a systematic approach to analyzing and gaining a better understanding of the ethical considerations of using AI for high-stakes decisions and ensuring fairness in using these systems.

Chain of Thinking

1. **Initial Parameters**
   "I need help with [clearly state your need].

2. **Process Guidance**
   To guide my thinking, provide a short list of questions related to [specific focus area].

3. **Additional parameters**
   These questions should help me expand and refine my thoughts on [reiterate focus area] and determine my next steps.

Sample Prompt: I need help with evaluating the ethical implications and fairness of using AI systems for high-stakes decisions, like student admissions or performance tracking. To guide my thinking, provide a short list of questions related to the potential biases, transparency, and accountability of these AI systems in educational settings. These questions should help me expand and refine my thoughts on ethical considerations in high-stakes AI decisions and determine my next steps for ensuring these systems are used fairly.

What do you notice and wonder about the results?

What connections can you make to this chapter?

Additional Task (Scenario Analysis):
Sunrise High School has started using an AI-powered admissions tool to help streamline the process of selecting students for its highly competitive Advanced Scholars Program. The AI reviews student applications, including grades, test scores, extracurricular activities, and teacher recommendations, to recommend who should be admitted.

The school administration relies heavily on the AI's recommendations, believing the system can eliminate bias and improve efficiency in the admissions process. However, a school counselor notices a troubling pattern after a year of using the system. Students from lower-income families and underrepresented racial/gender groups are being admitted to the program at much lower rates than their peers. When the counselor investigates, they learn that the AI system weighs standardized test scores and extracurricular activities more heavily, both favoring students from more affluent backgrounds. The counselor worries that the AI may perpetuate existing inequities by disproportionately disadvantaging students who don't have access to expensive test prep programs, have historically not been provided access to the program, or can't afford participation in certain extracurriculars.

As a result, the counselor brings these concerns to the school administration and recommends a review of the AI system's criteria. Analyze how the AI system in this scenario might be contributing to inequities in the admissions process. What are the ethical implications of using AI in high-stakes decisions like student admissions?

Discuss what steps the counselor and the school should take to ensure the AI admissions tool promotes fairness and equity. How might they adjust the criteria the AI uses to evaluate students?

Chapter 5 Concepts

Review the following key concepts in Chapter 5: *Synthetic Media, Deepfakes, "The Talk,"* and *Media Literacy.*

1. Identify and define any additional terms or concepts from this chapter that were new to you.

2. Reflect on your notes from this chapter in the book. Which concepts or ideas would you like to discuss further?

3. How do synthetic media, deepfakes, and media literacy concepts, including discussions like "The Talk," impact society beyond educational settings?

Chapter 5 Problems

Reflect on the following potential *perils* of AI in education and discuss them with others:

Bias Amplification: AI-generated educational content can inherit and amplify existing biases within the algorithms, potentially reinforcing harmful stereotypes and misrepresenting history. This could lead to discriminatory or inaccurate learning materials.

Psychological Harm: The rise of deepfakes poses a significant threat to students' mental health and well-being. False accusations, manipulated images, or videos targeting individuals or communities can lead to anxiety, self-doubt, and even trauma, especially for vulnerable students.

Erosion of Trust: Deepfakes make it increasingly difficult to discern real from fabricated content. This can undermine trust in educational materials, causing students to question the validity of information presented in the classroom and hindering their learning process.

Weaponized Deepfakes: Deepfakes can be weaponized to create false accusations against students or educators. This misuse of technology could lead to disciplinary action, reputational damage, and a hostile learning environment.

Classroom Disruption: Deepfakes can be used to disrupt the classroom by generating fake videos or audio recordings that spread misinformation, create conflict, and undermine the learning environment. This can challenge educators' ability to maintain order and a productive learning space.

Tasks:

Using the provided prompt, create a critical analysis of the use of synthetic media in a classroom setting.

Critical Analysis

1. **Define** the **scenario**
   Outline a specific, real-world example relevant to the classroom context.

2. **Identify** key **considerations**
   List and reflect on the issues inherent to the scenario.

3. **Reflect** and **respond**
   Ask questions to guide towards understanding the balance between application and responsibility.

4. **Conclude** with **actionable** ideas
   Feel free to ask clarifying questions if needed.

Sample Prompt: Imagine a scenario where students use AI-powered tools to create media projects for class. A group of students decides to use a deepfake tool to create a video of a historical figure or celebrity endorsing a modern issue. What ethical considerations should be addressed when students create synthetic media that portrays real figures? Consider the creative benefits and the risks of spreading misinformation or altering perceptions of real events. What guidance might an educator provide to ensure students understand the difference between creative expression and misrepresentation? How could you help students think critically about the ethical limits of using synthetic media? What specific guidelines could be set to promote ethical use of synthetic media in the classroom? Consider policies like adding disclaimers, including media literacy instruction, or setting boundaries on certain uses of synthetic media.

What do you notice and wonder about the results?

What connections can you make to this chapter?

Using an image generator, generate an image of all of the following:

1. Teacher
2. Gifted student
3. CEO
4. Doctor
5. Computer Scientist

Upon analyzing and considering the results, employ the following image generator prompt protocols to generate images with enhanced contextual depth and to mitigate the potential for bias in the results.

Image Generator

1. **Be Clear** - State exactly what you want
2. **Be Concise** - Use precise language without unnecessary words
3. **Provide Context** - Include setting, style, mood, or purpose
4. **Add Specific Details** - Describe important elements and characteristics
5. **Be Unambiguous** - Avoid vague or conflicting descriptions

What do you notice and wonder about the results? What was the main difference between the two activities?

What connections can you make to this chapter?

What connections can you make to Chapter 3?

Consider the modifications you would make to your prompt following the guidelines to mitigate the bias in the results.

Additional Task (Scenario Analysis):
In a teacher's social studies class, students are encouraged to use creative technology for their projects. This year, several students discovered an AI tool that allows them to create deepfake videos. The tool will enable them to insert historical figures into their projects to "speak" about events. One group of students created a video in which Abraham Lincoln gives a speech about modern social justice issues.

The students are excited by how lifelike the video is and how it adds a new dimension to their presentation. However, the teacher becomes concerned when another group of students presents a deepfake of a historical figure, Harriet Tubman, saying things she never actually said. The teacher realizes that while the technology has exciting potential for creativity, it also poses serious risks if students use it to spread misinformation, disinformation, or malinformation. The teacher wonders how to balance encouraging the creative use of AI with teaching students about the ethical risks and the importance of verifying information.

Analyze the potential benefits and risks of using deepfake technology in classroom projects. How can tools like these enhance learning, and what dangers do they pose regarding misinformation, disinformation, and historical accuracy?

Discuss the steps a teacher should take to ensure students use synthetic media responsibly. How can the teacher incorporate lessons on media literacy and critical thinking to mitigate the risks of deepfakes?

Finally, consider the ethical implications of using deepfake technology in education. How can educators ensure that the use of this technology aligns with ethical principles and avoids the potential for misuse?

Chapter 6: Student Data, AI Power

Chapter 6 Concepts

Review the following key concepts in Chapter 6:
Data Privacy, Informed Consent, Data Governance, and Data Literacy.

1. Identify and define any additional terms or concepts from this chapter that were new to you.

2. Reflect on your notes from this chapter in the book. Which concepts or ideas would you like to discuss further?

3. How do data privacy, informed consent, data governance, and data literacy affect students and families?

Chapter 6 Problems

Reflect on the following potential *perils* of AI in education and discuss them with others:

Transparency in Data Practices of Third-Party AI Tools: Schools often rely on third-party AI tools whose data collection methods are unclear. This lack of transparency makes it difficult for educators, parents, and students to understand how their data is used.

AI Bias in Educational Materials: AI algorithms are trained on existing data sets, which can perpetuate existing biases. These biases can then be reflected in the educational materials created using AI, potentially harming marginalized students.

Student Data for Unintended Purposes: Student data collected for educational purposes could be used for other purposes, such as targeted advertising, without the knowledge or consent of students or parents.

Student Control Over Data: Students often have little to no control over the data they generate through educational AI tools. This can lead to a sense of powerlessness and a need to understand how their data is used.

Data Security Risks: As more student data is collected and stored electronically, the risk of data breaches increases. This could expose sensitive student information, such as grades, disciplinary records, birthdates, addresses, and IEPs.

Chapter 6 Tasks

Tasks:

Using the provided prompt, create a controlled context analysis on the importance of data use and data privacy.

Context Control

1. **Give specific conditional references**
   When I use the word [term], I am referring to [specific meaning]. (e.g., When I say child, I mean student in grades 6-8.)

2. **Give a specific purpose**
   Your objective is to [clearly state the desired outcome].

3. **Give parameters**
   Focus your response strictly within the context defined by the clarified terms above.

Sample Prompt: When I use the term "data privacy," I am referring to protecting students' personal information from unauthorized access, misuse, or sharing without informed consent.

When I say "student data," I mean information collected through school systems, including academic performance, behavioral records, and any biometric data. Your objective is to create a framework for understanding the ethical responsibilities of handling student data within AI-powered educational tools. Focus your response strictly within the context defined by the clarified terms above, emphasizing data privacy protection, consent, and ethical data governance in educational AI tools.

What do you notice and wonder about the results?

What connections can you make to this chapter?

Additional Task (Scenario Analysis):
Sunset High School recently implemented an AI-powered platform to track student performance, attendance, and social-emotional well-being. The system collects vast amounts of student data, including their grades, behavioral reports, and school survey data. Student data collection complies with local, state, and federal regulations around data privacy, such as COPPA, FERPA, SOPIPA, and GDPR. The AI uses this data to

generate insights for teachers, identifying students at risk of falling behind academically or needing emotional support.

While the administration praises the AI for its ability to identify struggling students early, there is growing concern from parents about how much data is being collected on the child(ren) they care for attending the school. The worries include who has access to this information, how it's being used, whether the AI platform collects data beyond what is strictly necessary for educational purposes, and if any part could be shared with third parties.

Several parents are especially alarmed when they learn that the system uses facial recognition data from classroom cameras to track student engagement during lessons and campus cameras to track student behaviors. When one parent asks the school administration for a detailed explanation of how their child's data is being used and stored, they cannot provide a clear answer. This leaves the parent wondering whether the school has done enough to protect their child's privacy despite being compliant with data privacy regulations.

Analyze the potential privacy concerns raised using AI systems that collect extensive student data. What are the risks, and how might these systems affect student rights?

Discuss what steps schools should take to ensure transparency and protect student data. How can adult caregivers and educators advocate for stronger student data privacy protections? How might collecting facial recognition and other personal data violate or challenge data protection laws such as COPPA, FERPA, SOPIPA, or GDPR?

Chapter 7: Toward A More Equitable Classroom AI

Chapter 7 Concepts

Review the following key concepts in Chapter 7:
*Equity, Inclusion, Culturally Responsive Teaching,
and Design Thinking.*

1. Identify and define any additional terms or
    concepts from this chapter that were new to
    you.

2. Reflect on your notes from this chapter in
    the book. Which concepts or ideas would
    you like to discuss further?

3. How do equity, inclusion, culturally
    responsive teaching, and design thinking
    concepts relate to AI in schools?

Chapter 7 Problems

Reflect on the following potential *perils* of AI in education and discuss them with others:

Replicating Existing Inequities: Instead of transforming education, AI is often used to automate existing practices, potentially exacerbating existing inequities. This "remixing" of outdated methods fails to address fundamental flaws and may further marginalize vulnerable student groups.

Prioritizing Efficiency Over Equity: The focus on AI for efficiency and time-saving (e.g., automated grading) can overshadow its potential for promoting equity. This may lead to solutions prioritizing administrative convenience over addressing the root causes of educational disparities.

Neglecting Critical Pedagogy: AI tools are often implemented without considering critical pedagogy, which emphasizes challenging power dynamics and promoting social justice. This can lead to AI reinforcing existing hierarchies and failing to empower marginalized students.

Lack of Student Agency: Students are often positioned as passive recipients of AI tools rather than active participants in shaping their use. This lack of agency can limit AI's potential to promote critical thinking, creativity, and authentic learning experiences.

Insufficient Focus on Ethics and Equity: The development and implementation of AI in education often needs a strong focus on ethics and equity. This can result in biased algorithms, inadequate data privacy protections, and widening the digital divide.

Chapter 7 Tasks

Tasks:

Using the provided prompt to develop a thoughtful approach to using AI tools to promote culturally responsive teaching and equity in a classroom setting.

Reflection Pattern

1. **Give** it a **perspective**
   From your perspective, [pose an open-ended question or present a scenario].

2. **Process Parameters**
   When responding, always explain the reasoning behind your answer and clearly state any assumptions you're making.

3. **Challenge** the **Results**
   Throughout our conversation, suggest ways to improve my questions or prompts to get more helpful and relevant responses.

Sample Prompt: From your perspective as an educator, how could AI tools be designed to support culturally responsive teaching and promote equity in a diverse classroom setting? When responding, always explain the reasoning behind your answer and clearly state any assumptions you're making, such as assumptions about the student's needs, the school's resources, or the role of AI in education. Throughout our conversation, suggest ways to improve my questions or prompts to get more insightful and relevant responses. For example, recommend alternative ways to phrase questions to understand better how AI can address specific equity challenges or engage diverse student backgrounds effectively.

What do you notice and wonder about the results?

What connections can you make to this chapter?

Additional Task (Scenario Analysis):
At Brookstream High School, the administration has recently introduced AI-powered tools to streamline several educational tasks. One of these tools, based on significant demand and volume of requests from teachers, is an AI plagiarism detector that scans students' written assignments to identify potential instances of AI-generated content or plagiarism. Teachers are encouraged to use the AI detector to ensure academic integrity. The students are even

provided guidance on how to use AI in their handbook. In some instances, teachers determine when it is permitted to be used and when it is not permitted to be used.

Initially, many teachers appreciate how the detector quickly flags assignments that may require closer inspection. However, several start to notice that certain groups of students, particularly those for whom English is a second/additional language or who come from marginalized backgrounds, are more frequently flagged by the AI. When these teachers investigate, they find that the AI incorrectly flags students for "plagiarism" based on common phrases or non-standard grammar, even when the work is original. This group of teachers worry that the AI detector may be biased against students with non-traditional writing styles and could unfairly penalize them.

Furthermore, the teachers grow concerned when they discover that the AI detector collects and stores a large amount of student data beyond what is necessary for "plagiarism" detection. They suspect that the data may be used to train the AI platform further, shared with third-party companies, and potentially used for targeted advertising without students' or parents' knowledge or consent, potentially violating specific data protection laws.

As the school relies more heavily on this AI detector and other automated systems, this group of teachers also noticed a shift in student learning. Students seem more focused on avoiding detection by the AI rather than engaging deeply with their writing. This over-reliance on automation has led to shallow learning, as students are more concerned with producing work that passes the AI detector rather than improving their critical thinking and writing skills.

Meanwhile, this group of teachers also feels that their role as teachers is being diminished. Instead of guiding students through the writing process, they spend more time managing the AI detector's reports and resolving false flags. They worry that AI tools devalue her expertise and reduce their ability to foster meaningful learning experiences.

They find little transparency when they try to understand how the AI detector makes its decisions. The system's inner workings are opaque, and they cannot challenge the results or address potential biases in the algorithm. This lack of explainability makes it difficult for them to trust the AI tool and leaves them questioning its fairness and reliability.

Analyze the ethical concerns raised by the teachers regarding the AI detector. How might issues such as biased flagging, excessive data collection, and potential privacy violations compromise the ethical use of AI in the classroom? How could the AI detector disproportionately flag students from marginalized backgrounds?

Discuss the impact of over-reliance on AI tools like plagiarism detectors. How could this reliance affect students' learning experiences and their development of critical thinking skills? What can be done to promote deeper learning while using AI tools? Consider the role of transparency and explainability in AI systems. What should teachers and administrators demand from AI developers to ensure the system is explainable and accountable? What steps could the school take to address these concerns and ensure fairness?

Finally, consider the potential for AI to be used to create personalized learning experiences. How can AI tools be used to tailor instruction to the individual needs of each student? What are the ethical implications of using AI in this way?

Conclusion

As we embrace AI's potential to revolutionize education, we must equally commit to addressing its ethical challenges, including issues of equity, bias, learning impact, data governance, and student privacy. Our mission as educators is to harness AI not as a replacement for human judgment but as a tool to enhance our teaching, empower diverse learners, and actively dismantle systemic inequities in education.

www.ingramcontent.com/pod-product-compliance
Lightning Source LLC
Chambersburg PA
CBHW060944120626
46557CB00003B/1139